# GOLDEN
## 1941-1945
### (USN/USMC Aircraft of World War II)

## by Jim Sullivan & Dave Lucabaugh
### Illustrated by Don Greer

**squadron/signal publications**

A Marine PBJ-1J of VMB-611 makes a low pass over his South Pacific home base during 1945. Marine Mitchell bombers were widely used in the island hopping campaign.

COPYRIGHT © 1993 SQUADRON/SIGNAL PUBLICATIONS, INC.
1115 CROWLEY DRIVE   CARROLLTON, TEXAS 75011-5010
All rights reserved. No part of this publication may be reproduced, stored in a retrieval system or transmitted in any form by any means electrical, mechanical or otherwise, without written permission of the publisher.

ISBN 0-89747-294-2

If you have any photographs of the aircraft, armor, soldiers or ships of any nation, particularly wartime snapshots, why not share them with us and help make Squadron/Signal's books all the more interesting and complete in the future. Any photographs sent to us will be copied and the original returned. The donor will be fully credited for any photos used. Please send them to:

Squadron/Signal Publications, Inc.
1115 Crowley Drive
Carrollton, TX 75011-5010

# Acknowledgments

We would like to thank everyone who assisted in the preparation of this effort for their outstanding cooperation in making material available from their collections. A special thank you is due to Bill Larkins, Bill Derby, Jr., Stan Piet, Dave Ostrowski and Roger Besecker for their "extra-effort" in locating needed photographic material. Our sincere thanks to all.

| | |
|---|---|
| Harold Andrews | Earl Neff |
| Roger Besecker | Dave Ostrowski |
| Pete Bowers | Bruce Porter |
| Bill Curry | Stan Piet |
| Lou Darden | Ray Pritchard |
| Larry Davis | W. E. Scarborough |
| Bill Derby, Jr. | Bob Searles |
| National Archives | W. F. Gemeinhardt |
| Ted Stone | Clay Jansson |
| Hank Weimer | Duane Kasulka |
| Jim Wiedie | Bill Larkins |
| U. S. Marine Corps | Paul McDaniel |
| NASM-Smithsonian Institution | U. S. Navy |
| Joe Michaels (J.E.M. Slides) | Harry Gann/McDonnell-Douglas |

# Dedication

For all the brave pilots and aircrewmen who served the Navy and Marines during the raging conflict known as the Second World War. We salute all who safely returned and remember in grateful appreciation those who paid the ultimate price to protect our liberty and freedom — heros all!

(Overleaf) Navy and Marine aircraft line the airstrip on Majuro Island during March of 1944, including F6F-3s of VF-39, as well as SBDs, TBFs, PBYs and R4Ds. Aircraft were often towed around the field whenever possible to hold down the clouds of coral sand that were thrown up by the propeller blast. (National Archives)

This SB2U-2 (BuNo 1369) of VS-72 on the ramp at Chambers Field, NAS Norfolk, VA during January of 1941 displays the colorful pre-war color scheme of Natural Metal with Yellow wings and colored tails that decorated Naval aircraft of the period. Vindicators served with both the Navy and Marines. This aircraft carries the squadron insignia and battle-efficiency "E" award on the fuselage. (W. E. Scarborough)

# INTRODUCTION

As a result of the massive surprise Japanese bombing attack at Pearl Harbor on the morning of 7 December 1941, the United States of America entered the Second World War. Although the distant rumblings of war had been heard months earlier, the U.S. military was not fully prepared for combat on a worldwide scale. Our military aircraft were relatively few in numbers and consisted of trainers, transports, scouts, fighters and bombers. Many of these were obsolete when compared to the performance of the Mitsubishi Zero that formed the backbone of the Japanese Empire's aerial assault.

Fortunately for the United States, the few aircraft carriers we had in the Pacific at the time of the Pearl Harbor attack were at sea and sustained no damage as a result of the Japanese aerial assault. The USS ENTERPRISE was enroute back to Pearl and its air group suffered limited losses of SDBs and F4Fs as they arrived in the midst of the enemy air raid. This fact alone allowed the U. S. to preserve enough Naval air power to rebuild and reorganize. Starting the day after Pearl Harbor, a rapidly accelerated buildup of men and material began, which did not let up or level off until near the war's end in September of 1945. This book pictorially depicts this rapid development and the scope of aircraft utilized by the United States Navy and Marine Corps from 1941 thru 1945.

The Second World War saw many aircraft types used, including the star performers, the F6F Hellcat and the F4U Corsair. These fighters, however, were just two of the types used by the Navy and Marines. Other aircraft, although lesser known, provided vital service that was essential to the ultimate result of the war effort. Victory was attained by the combined results of all aircraft types such as trainers, flying ambulances, cargo haulers, scouting planes, target tows, fighters and bombers. There were long-range patrol aircraft that flew for countless monotonous hours to keep the sea lanes open and relatively free of enemy subs. The photo reconnaissance aircraft braved danger daily to bring back vital photographic information for upcoming raids as well as post strike photographs to determine the damage done by these strikes.

Under primitive conditions, many of the land-based pilots lived and flew from airstrips that were not much more than muddy fields. As the war progressed, conditions under which both men and machines were forced to operate improved, thanks to the arrival of the Navy Seabee and Army construction teams. Muddy strips became operational airfields and living conditions were upgraded from tents to Quonset huts. Aircraft maintenance personnel were considerably better housed and equipped to keep the aircraft flying.

As American industry worked feverishly to bring new planes from the drawing board to the battlefield, our air power gradually turned the tide of Japanese air superiority. Once air superiority was established, the island-to-island war became a winnable situation as one-by-one, the Japanese held islands fell and the military forces of the United States inched ever closer to the home islands of Japan. Air power was the key that enabled our fighting forces to ultimately prevail over the Empire of the Rising Sun.

**An undamaged PBY-5 of VP-12 that survived the Japanese attack on NAS Kaneohe on 7 December 1941 and the still burning Catalina in the background are quite a contrast. Only four of the squadron's twelve PBY-5 Catalinas survived the Japanese attack. (National Archives)**

**J2F-3s and an RD-3 (BuNo 9532) on the ramp outside hangar 37 at NAS Ford Island on 7 December 1941 just after the second wave of the Japanese attack. In the hangar are a pair of SBDs and a Vought OS2U Kingfisher. (National Archives)**

**With its cockpit cover still in place, this Grumman J2F-4 of VJ-2 was one of the lucky Pearl Harbor survivors on NAS Ford Island. It soldiered on with various units until it was stricken from the inventory during February of 1945. VJ-2s tail color was Lemon Yellow. The PBY-1 in the background was also assigned to VJ-2. (National Archives)**

# 1941

The Navy and Marines were suddenly confronted with the opening of hostilities on that Sunday morning of 1941 by a series of devastating air attacks from Japanese carrier-based bombers. Our Hawaiian bases were served by a wide variety of aircraft types. Unfortunately, none of these could truly have been considered "World Class" combat machines when compared to the fighters and bombers in service in the European air war. It became quickly evident that the little known Japanese Zero fighter was superior to our own Wildcat and Buffalo carrier-based fighters.

The compiled report on the Status of U. S. Naval Aircraft on 6 December 41 stated that there were 2,471 combat types, 2,459 training types and 303 utility types, for a grand total of 5,233 aircraft on hand. This total would continue to increase dramatically over the next forty-five months until the climax of the Second World War on VJ-Day in 1945.

Fortunately, the aircraft destroyed at NAS Kaneohe, NAS Ford Island and MCAS Ewa during the Japanese attacks were primarily land-based scout and patrol types rather than the more critically needed carrier based fighters, torpedo bombers and scouts which were in short supply. The results of the Japanese bombing attack saw the destruction of many PBY, OS2U and SOC aircraft on NAS Ford Island and Kaneohe while the few carrier type planes that were strafed and burned at MCAS Ewa were part of the Marine Air Group. Even though the aircraft lost were important to the naval air mission, a point could be made that these non-carrier machines were more readily replaceable from the stateside aircraft pools and East coast squadrons. At that time, and overall, it was more important to have the Carrier Air Groups remain relatively intact. Only the Enterprise Carrier Air Group suffered minor losses due to their untimely arrival at Pearl Harbor in the midst of the enemy attack.

**On the day after the Pearl Harbor attack aircraft that survived or were salvagable were gathered together on Ford Island. The ramp in front of Hangar 38 was packed with surviving PBY Catalina, OS2U Kingfisher and SOC Seagull aircraft. (National Archives)**

At dawn on Monday, 8 December 1941, the first order of wartime commitment for Navy Air was the mounting of a far-reaching search from Hawaii in a concentrated effort to locate the Japanese Fleet that had brought destruction to the ships, aircraft, facilities and personnel at Pearl Harbor. The frantic search by the surviving Navy planes proved to be totally fruitless as the enemy fleet had withdrawn and was not to be found.

The regrouping process and the replacement of lost aircraft continued in the Pacific Fleet through the remaining twenty-three days of 1941.

**Seven PBY Catalinas are parked on the ramp in front of Partol Wing Two's burned-out hangar on NAS Ford Island on 8 December 1941. The battleship in the background is the heavily damaged USS NEVADA which was beached to avoid sinking. (National Archives)**

(Above) This Consolidated PBY-5 Catalina on the ramp at NAS Kaneohe on 8 December 1941 was heavily damaged in the Japanese attack with most of the wing fabric burned off, the port pontoon shorn off and burned metal on the forward fuselage. (National Archives)

(Below) The prototype Vought XTBU-1 (BuNo 2542) sits on the Vought factory ramp at Stratford, CT on 20 December 1941. Although the Seawolf was designed by Vought, the 180 production aircraft were turned out by Consolidated Vultee under the designation TBY. (Vought)

Vought OS2U-1 Kingfishers of VS-2D1, an Inshore Patrol Squadron, are tucked safely inside Seaplane Hangar #2 at NAS Quonset Point, RI on 27 February 1941. The second Kingfisher has the engine cowling removed for routine maintenance work. (National Archives via Bill Curry)

This F4F-3 (BuNo 2527) of VF-42 on the taxiway at NAS Norfolk, VA during March of 1941 has the tail surfaces in Willow Green and the wings in Chrome Yellow, while its squadron mate is in overall Light Gray camouflage. (W. E. Scarborough)

This Curtiss SOC-1 Seagull of VCS-6 is configured with its optional wheeled landing gear. The Seagull normally operated on floats, serving as scout-observation aircraft on cruisers. The aircraft is overall Light Gray with White codes. (Pete Bowers)

This PBY-5 of VP-52 was one of the first Catalinas of the first squadron to arrive at the brand new, still under construction, Naval Air Station at Quonset Point, RI on 26 March 1941. (Navy via Bill Curry)

This Brewster F2A-2 Buffalo (BuNo 1403) on the ramp at Oakland Airport, CA on 28 October 1941 carried large Black numbers on the fuselage for easy recognition. The overall Light Gray aircraft was assigned to the Advanced Carrier Training Unit at NAS North Island, San Diego, CA. (Pete Bowers)

An OS2U-1 (BuNo 1683) of VO-1, assigned to the battleship USS ARIZONA, taxies up to the ship during recovery operations off Hawaii on 6 September 1941. This Kingfisher was lost during the Japanese attack on Pearl Harbor. (National Archives)

A factory fresh Curtiss SNC-1 Falcon warms up prior to starting another training mission during 1941. The Falcon served as an advanced combat trainer, supplementing the SNJ, although it was produced in far fewer numbers. (Bill Larkins via W. F. Gemeinhardt)

This Naval Aircraft Factory SBN-1 (BuNo 1538) of VT-8 crashed after hitting a drainage ditch while attempting to land at NAS Hampton Roads, VA on 4 October 1941. VT-8 was the only operational fleet unit to use the SBN-1. They flew them for a short period before being re-equipped with the Douglas TBD-1. (National Archives)

A Vought SB2U-3 (BuNo 2075) Vindicator of VMSB-131 balanced on its prop after nosing over at Mustin Field, PA (near the Naval Aircraft Factory in Philadelphia) on 21 May 1941. This Vindicator, one of fifty seven built, remained on active service thru mid-1943. (National Archives)

An overall Light Gray Wildcat of VF-71, awaits clearance to take the runway at Chambers Field, NAS Norfolk, VA during January of 1941. Standard armament for this version of the Wildcat consisted of four .50 caliber machine guns. (W.E. Scarborough)

These Douglas TBD-1 Devastators of Torpedo Squadron Six (VT-6) were gathered together on the ramp at NAS North Island, San Diego, CA during the Summer of 1941 for the filming of the classic Naval Aviation movie, "DIVE BOMBER." (National Archives)

The Stearman N2S-2 was the standard primary trainer for the Navy giving fledgling flyers their first exposure to military flying. This N2S-2 formation belonged to VN-11D8 flying out of Rodd Field at NAS Corpus Christi, TX during November of 1942 (National Archives)

# 1942

The beginning of 1942 saw a massive buildup begin for the Navy and Marine units in the Pacific as they recovered from the aircraft losses at Wake Island and the Philippines, with the continued movements of squadrons and aircraft from the states to Hawaii and beyond.

The Pacific Fleet Carrier Air Groups were basically kept at sea due to a decision to have only one carrier in port at Pearl Harbor at any given time. This dispersal was a direct result of the lesson painfully learned on 7 December.

No significant contact was made with the Japanese carrier units until February of 1942 when the carriers made their first raids on enemy held islands. The first major carrier combat took place in May of 1942 in the Battle of the Coral Sea which resulted in the loss of the first U. S. carrier, the USS LEXINGTON, and most of her Air Group. This was followed shortly by the pivotal encounter at Midway which saw the first operational use of a new Navy type, the Grumman TBF-1. The Avenger torpedo-bomber was the intended replacement for the obsolete TBD-1 Devastator. In spite of the TBF's poor initial combat record at Midway with VT-8, the Avenger went on to become the sole operational carrier VT aircraft for the remainder of the Second World War.

Throughout the first part of 1942, and culminating with the Battle of Midway which ended on 4 June, the one aircraft that emerged as outstanding was the very reliable, and most effective, Douglas SBD. The Dauntless had shown itself to be a superior dive bomber, perhaps the best in the world against ship targets. It proved to be the single pre-Pearl Harbor Navy type capable of functioning as both a carrier or land-based scout/dive bomber past mid-1944 — quite a remarkable record!

After Midway, there was somewhat of a lull in naval air encounters until August of 1942 when the deadly and delicately-balanced aerial struggle for Guadalcanal began. By that time, the Brewster F2A Buffalo, Douglas TBD-1 and Vought SB2U-3 had been removed from combat service. They had been declared obsolete and were replaced by newer aircraft. The Grumman F4F-4, TBF-1 and the Douglas SBD-5 which would now engage the enemy. The cruel test of combat over Guadalcanal would severely test the rugged F4F-4 Wildcats. In the capable and aggressive hands of USMC fighter pilots, Wildcats bore the brunt of fighter combat for the remainder of 1942 until the island was finally secured. During this time at sea, there were several major encounters which resulted in the loss of two carriers, the USS HORNET and USS WASP. As the year progressed past mid-point, the great American production capability began to take hold and show results. We easily replaced aircraft combat losses and also provided new aircraft for training use. Additionally, the aircraft factories working around the clock produced aircraft to fill the needs of new squadrons being formed to man the new carriers under construction as well as the increasing number of scouting and patrol squadrons being put into service to meet the far-flung commitments of USN/USMC air power in the wide expanses of the Pacific Theater.

A Vought SB2U-2 Vindicator of Scouting Squadron Nine (VS-9) settles down on the flight deck of the training carrier USS CHARGER On 29 October 1942. The Vindicator was camouflaged Intermediate Blue-Gray over Light Gray with Black codes. (National Archives)

This Douglas RD-4 Dolphin (V-128) was assigned to the San Francisco Coast Guard Station, CA during June of 1942 and carried perhaps the largest national insignia carried on any Second World War naval aircraft. The beaching gear was detachable and removed after the seaplane was afloat. (Bill Larkins)

This Sikorsky JR2S-2 (BuNo 12390), moored off NAS Banana River, FL on 4 June 1942, was one of three which served the Navy during the Second World War. It was impressed into military service from American Export Airlines and was the largest seaplane to see Naval service during the war. (National Archives)

This was the first PBY-5 Catalina (V189) to be assigned to the U.S. Coast Guard. It was stationed at the San Francisco Coast Guard station on 3 June 1942. Once the beaching gear was attached, the aircraft was pulled up onto the seaplane ramp at the USCG Station. (Bill Larkins)

TBD-1s and F4Fs are tightly spotted on the flight deck of USS ENTERPRISE (CV 6) on 11 April 1942. The Devastator carrying the side number 8 was BuNo 0387, assigned to Torpedo Squadron Six (VT-6). TBDs first joined the Navy in late 1937 and saw combat against the Japanese in early 1942. (National Archives)

A Grumman TBF-1 Avenger banks gently over the Atlantic just off the Virginia coast during 1942. The TBF had a very successful career flying with both the Navy and Marine Corps. It was powered by a 1,700 hp Wright R-2600-8 air cooled radial engine giving it a top speed of 271 mph. (National Archives)

This Vought F4U-1 (BuNo 02296) was the 144th Corsair to roll off the Vought production line at Stratford, CT during April of 1942. Its first squadron was VMBF-931 at MCAS Cherry Point, NC. It survived the Second World War and was stricken from the inventory in late November of 1945. (National Archives)

A flight of OS2U-3 Kingfishers (13 - BuNo 5865, 17 - BuNo 5869 and 16 - BuNo 5868) of VS-1D1, an Inshore Patrol Squadron, fly in line abreast formation. The rear facing spike on the front underside of the main pontoon was used to snag the recovery sled towed by cruisers and battleships. These OS2U-3s were based at NAS Squantum, MA during May of 1942. (National Archives)

This Beech SNB-1 (BuNo 39800) served with Fleet Air Wing Twelve (FAW-12) at NAS Norfolk, VA during 1942 and was used primarily for pilot and gunnery training. The clear nose cap, twin .50 caliber machine gun turret and long fuselage fairings were modifications to the basic Kansan airframe. (Beech via Pete Bowers)

A Douglas SBD Dauntless of VS-41 catches a wire aboard the USS RANGER during 1942. The Dauntless was flown by both the Navy and Marines, serving on land bases as well as carriers. Marine SBDs were still in active service during the Philippine Invasion as close air support aircraft. (National Archives)

This Lockheed JO-2 (BuNo 1049) was a six-seat transport assigned to Headquarters, U.S. Marine Corps in Washington, DC and was based at NAS Anacostia during early 1942. It was one of three that served with the Marines. (Pete Bowers)

SNC-1 Falcons were rarely camouflaged. This Intermediate Blue-Gray over Light Gray aircraft carries the national insignia in six positions and has the prop tip warning stripes in Red/Yellow/Black. The mission of this Curtiss-built, two-place monoplane was advanced pilot training and it was based at NAS Daytona Beach, FL during late 1942. (National Archives)

TBD-1s of Torpedo Squadron Four (VT-4) and F4F-4s of Fighting Squadron Forty-one (VF-41) along with Vought SB2Us tightly pack the flight deck of USS RANGER on 18 June 1942. The folded wings of the Devastators efficiently kept the height of the TBD-1 to a minimum. (National Archives)

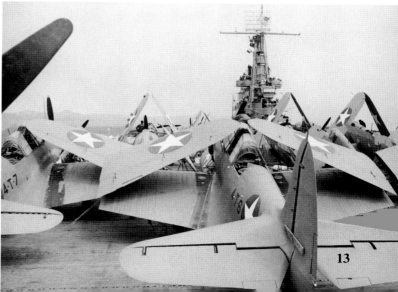

This North American SNJ-3C (BuNo 01876) served with the Carrier Qualification Unit at NAS Glenview, IL during late 1942. It was overall Natural Metal with Black lettering and wide Green wing stripes. It was operating from USS WOLVERINE, one of two fresh water training carriers stationed on the Great Lakes. (National Archives)

This Piper HE-1 (BuNo 30198), stationed at NAS Corpus Christi, TX during December of 1942, was a flying ambulance capable of carrying one litter patient. The upper part of the fuselage behind the wing swung upward, much like the trunk of a car, to accomodate the litter. Special ambulance markings were applied in the form of a Red cross on a circular White field. (National Archives)

A Naval Aircraft Factory N3N-3 (BuNo 2928) in the hangar at Naval Auxiliary Air Station (NAAS) Rodd Field, TX during November of 1942, with its engine accessory panels removed for maintenance. The N3N-3 served as one of the Navy's primary trainers or "Yellow Perils" during the war. It was very agile, a delight to fly and had a top speed of 125 mph. (National Archives)

The Timm N2T-1 (BuNo 39188) was a two-seat primary trainer made of all plywood construction. The Tutor had a gross weight of just over 2,700 pounds and could reach a top speed of 144 mph. This rarely-seen Navy trainer carried fuselage bands and large side numbers that were unique to Corpus Christi, TX area fields in November of 1942 (National Archives)

(Above) A Lockheed PBO-1 (BuNo 03854) of VP-82 rests on its belly after a wheels-up landing at the Naval Aircraft Factory, Mustin Field, PA on 12 January 1942. Built as part of a batch of Hudsons destined for England, twenty were reassigned to the Navy. They retained their original British camouflage with the addition of U.S. insignia and rudder stripes. (National Archives)

(Below) A TBF-1 of Torpedo Squadron Eight (VT-8), fresh from Grumman's Long Island factory, on the ramp at NAS Norfolk, VA on 23 March 1942. The thirteen Red and White alternating stripes were short-lived as they were soon removed by the squadron. The aircraft was camouflaged Intermediate Blue-Gray over Light Gray with Black lettering. (National Archives)

This Curtiss SO3C-1 (BuNo 4876) has just been recovered by the aircraft crane aboard the light cruiser USS COLUMBIA on 8 November 1942. This Curtiss design proved to be totally unfit for this mission and was soon removed from all ship-based operations and replaced by the reliable SOC biplane. (National Archives)

This Curtiss SBC-4 of VMO-151 was one of the few pre-war biplane types that saw combat during the early part of the Second World War. A single 500 pound bomb was loaded beneath this Marine Helldiver for anti-submarine patrols off the coast of American Samoa during early 1942. (via Bruce Porter)

An F4F-4 of VGS-1 is position forward after landing aboard the escort carrier USS LONG ISLAND on 17 June 1942. Since recovery operations were still going on, the barrier wires were raised to protect the parked aircraft spotted forward. (National Archives)

This Sikorsky JRS-1 (side number 5), on the ramp at NAS Ford Island Hawaii during the Spring of 1942, carried depth charges on the underwing bomb racks located outboard the engine nacelles. This amphibian was one of the utility types pressed into combat services out of Hawaii due to the heavy losses of patrol aircraft at Pearl Harbor on 7 December. (National Archives)

Grumman F4F-3 Wildcats of VF-8 and several Curtiss SBC-4 Helldivers of VS-8 parked on the forward flight deck of USS HORNET, moored at the Norfolk Navy Yard in Portsmouth, VA on 28 February 1942. (National Archives)

A Douglas SBD-3 of VGS-28 comes to an abrupt halt after engaging the arresting cable on the excort carrier USS CHENANGO (CVE-28) during December of 1942. The aircraft was camouflaged Intermediate Blue-Gray over Light Gray with Black fuselage lettering. (National Archives)

A Curtiss SOC-3A Seagull of VGS-30 lifts off the flight deck of USS CHARGER on 10 May 1942. This version of the Seagull was equipped with an arresting hook for carrier operations and participated in action throughout the Second World War. (National Archives)

These North American SNJ-3s (109 - BuNo 01947 and 107 - BuNo 01927), on the ramp at NAAS Kingsville, TX during November of 1942, belonged to the Training Command. The Texans were overall Natural Metal with wide Yellow wing bands and Black lettering. (National Archives)

A Martin PBM-1 Mariner of VP-74 firmly secured to the ramp at Darrell's Island, Bermuda during March of 1942. This seaplane had beaching gear installed and featured retractable wing floats. The dark spots on the side of the fuselage were ASW radar antennas. (National Archives)

# 1943

A flight of Douglas SBD-5 Dauntless dive bombers of VB-16 fly in formation while returning from a mission against Japanese-held Mille Atoll on 22 November 1943. The large fuselage national insignias were common on SBDs. (National Archives)

1943 marked the continuing American effort to build up Naval air power to meet the far-flung commitments in the Pacific area plus controlling the deadly submarine threat in the Atlantic. To meet these objectives, new aircraft carriers were started along with an enormous training program to staff new operational units. 1943 saw astounding increases in the production of a wide variety of aircraft types required to fill these urgent needs.

This year saw the operational debut of a new carrier-based fighter, the Grumman F6F-3. The Hellcat would eventually become the backbone of carrier-borne fighter squadrons throughout the remainder of the war. The F6F-3 made its first combat appearance on 31 August 1943 with VF-5 and VF-9, flying from the USS YORKTOWN and USS ESSEX respectively, during the first raids on Marcus Island. The Hellcat performed very well and soon completely replaced the Wildcat on all carriers except escort class ships. Only the General Motors FM-2 Wildcat continued to operate from these smaller carrier decks. The earlier Grumman-produced F4Fs were re-assigned to the Training Command to help new fighter pilots learn and hone their skills.

Another hard-hitting fighter also made its combat debut in 1943, the Vought F4U. The Corsair first tasted combat on Guadalcanal with VMF-124 during mid-February. It came into wide operational use only after a long development period. As a carrier-based fighter, it was assigned to VF-17 for its initial carrier deployment in July of 1943. Unfortunately, the F4U had several shortcomings in routine flight deck operations and was withdrawn from the carriers. What was a disappointing turn of events for carrier aviation proved to be a blessing for the Marines, since the Corsair was turned over to the Corps to operate from land bases. Almost immediately, the Corsair displayed an ability to effectively deal with anything the Japanese could put into the air. Vought's star performer, in several different variants, continued on as the Marine's frontline fighter for the remainder of the Pacific war.

Still, a third important aircraft also made its operational appearance during 1943, the Curtiss SB2C-1 Helldiver. The Helldiver had a very lengthy development cycle, from the prototype's first flight to full deployment aboard fleet carriers. Unsuccessful early carrier trials aboard the USS ESSEX in November of 1942 showed that further refinements were needed before the Helldiver could go to the fleet. It was delayed from carrier duties until an extensive test and modification program was completed. This program was concluded by May of 1943 and the reworked SB2C-1 was assigned to the carrier USS YORKTOWN.

Following extensive carrier flight operations, the reworked Helldivers were still deemed unacceptable for the Fleet use by the USS YORKTOWN's skipper and he had them removed from his ship and replaced by the more reliable SBD-5 Dauntless. It was November of 1943 before the long-suffering SB2C-1 finally was released to the Fleet and entered combat. Its first combat occurred on 11 November 43 when Helldivers, flying from the deck of the USS BUNKER HILL, attacked Rabaul. The big Curtiss design had finally "arrived" and assumed its role as the workhorse naval dive bomber. It served both the Navy and Marines for the remaining twenty-two months of the Pacific conflict.

As the second full year of the war ended, the buildup of Navy and Marine Air had still not reached its peak, but it did have all of the major new aircraft types in place and available in quantity along with an ample number of well-trained pilots and air crewmen, representing the largest naval air arm in the world.

A Lockheed PV-1 of VB-136 on patrol off the Alaskan coast on 1 May 1943. This Ventura carried the national insignia in eight positions. The marking on the rear fuselage was a smiling bomb with the message *Right In TOJO's Face*. (National Archives)

An early production Vought F4U-1 Corsair with bomb racks installed under the wing just outboard of the .50 caliber machine gun ports. The pilot was warming up the 2,000 hp P&W R-2800 engine prior to launch from the USS CHARGER on 8 March 1943. (National Archives)

An SOC-3 of VCS-6 being hoisted aboard the heavy cruiser USS MINNEAPOLIS (CA-36) after making a safe water landing while operating off Wake Island on 5 October 1943. The Seagull would be positioned either on the catapult or hoisted below deck for storage in the stern hangar for maintenance. (National Archives)

This PBM-3 Mariner, flying out of the large PBM crew training base at Banana River, FL during August of 1943, had recently changed its markings from the earlier star/circle insignia to the star/bar insignia with a Red surround. This Mariner variant had fixed wing floats and a radome over the fuselage. (National Archives)

This Waco VKS-7 (BuNo 37649) carries a variety of mixed markings and was assigned to Pan American Airways who was a cargo carrier operating under the Naval Air Transport System (NATS). The aircraft was later impressed into the Navy for pilot training at Concord, CA during October of 1943. (Bill Larkins)

This Marine SNJ-3 Texan over Turtle Bay airstrip on the island of Espiritu Santo was attached to MAG-11 during October of 1943. Only a small number of SNJ Texans made it to the South Pacific combat area. The aircraft was Intermediate Blue-Gray over Light Gray. (National Archives)

Marine Technical Sergeant J.A. Watson installs .30 caliber ammunition belts around the underwing 100 pound bomb in a Marine Corps SBD-3 Dauntless. This Marine field adaptation was known as a "daisy cutter" and was used as an anti-personnel weapon. (National Archives)

This Marine Curtiss R5C-1 Commando on an island base in the South Pacific on 28 November 1943 was camouflaged with Olive Drab uppersurfaces over Neutral Gray undersurfaces. Some 160 R5Cs were flown by the Marines who considered it a great freight-hauler and transport. (USMC)

(Right) This SOC-1 (BuNo 9941) of VCS-7 has snagged the recovery sled and in short order was hoisted aboard the heavy cruiser USS TUSCALOOSA on 19 July 1943. The rear seat gunner moved into position to make the hook-up of the Seagull onto the ships crane. (National Archives)

A Marine aviation ordnanceman checks the bomb load of a Marine SBD-3 (151) on Bougainville Island, Solomon Islands Chain on 20 December 1943. The SBD-3 carried a fuselage mounted 1,000 pound bomb and two 100 pound bombs on wing racks. The Marine Corps operated the Dauntless exclusively from land bases. (National Archives)

(Above) Deck crews spread the wings of an F6F-3 Hellcat of VF-9 aboard USS ESSEX on 5 April 1943. Only very early production F6F-3 Hellcats had gun fairings over the barrels of their six .50 caliber Browning machine guns. (National Archives)

(Below) A TBF-1 of VC-30 turns up on the flight deck of USS MONTEREY just prior to launch on 28 November 1943. This Avenger carried White lettering on the fuselage and a Red surround national insignia in the standard four positions. (Hank Weimer)

With its wings folded, an F6F-3 (side number 00) of VF-5 is positioned on the deck-edge elevator aboard the USS YORKTOWN on 6 May 1943. This Hellcat was flown by the Air Group Commander, CDR James Flatley, and is from the early production run with wing gun fairings. (National Archives)

An Olive Drab over Neutral Gray Douglas R5D-1 Skymaster of VR-5 on the dirt ramp of Adak, Alaska during August of 1943. Serving with both the Navy and Marines, the R5D-1 was capable of hauling passengers or cargo, was powered by four P&W R-2000-7 engines and had a top speed of 281 mph. (National Archives)

A PBM-3S (BuNo 01679) of VP-209 cruises over the Atlantic off the coast of Virginia near NAS Norfolk on 17 August 1943. The large dome just behind the cockpit housed surface search radar. The Martin Company delivered 156 of the PBM-3S versions to the Navy. (National Archives)

The Vought XTBU-1 (BuNo 2542) was assigned as a radio test plane at NATC Patuxent River, MD during late 1943. This was the only Seawolf produced by Vought since their commitment to the F4U-1 Corsair program tied up their production capability. As a result, Consolidated-Vultee assumed the Seawolf production contract. (National Archives)

This Douglas BD-2 (BuNo 7040) was a Marine utility aircraft assigned to target and glider towing duties at Page Field, Paris Island, SC during 1943. The Havoc carried the name *Lil Abner* on the nose in White along with a painting of the cartoon character. (W. F. Gemeinhardt)

This Consolidated PB4Y-1 (BuNo 32198) of VPB-114 carried the Atlantic sea-search paint scheme of Dark-Gray over White during December of 1943. Liberators routinely flew long anti-submarine patrols from coastal bases such as NAS Norfolk, VA. (National Archives)

(Above) An SBD-4 (BuNo 06947) of VC-22 on final approach for landing aboard the USS INDEPENDENCE on 1 May 1943. Canopies were normally kept open during takeoff and landing so that the crew could make a speedy exit from the aircraft in the event of a mishap. (National Archives)

(Below) This Curtiss SB2C-1 Helldiver (BuNo 00020) was serving with Bombing Five aboard the USS YORKTOWN (CV-10) on 7 May 1943. The extended slats on the wing leading edge increased the lift and resulted in lower takeoff and landing speeds. (National Archives)

The Eastern Division of General Motors produced over 4,400 FM-2 Wildcats for the Navy at Eastern's Linden, NJ facility. The FM-2 was a tall-tail version of the Wildcat which continued in service throughout the Second World War. (Navy)

A pair of R4D-1s of SCAT (South Pacific Combat Air Transport) fly in echelon formation toward Bougainville Island during March of 1943. Most Douglas Skytrains retained their Army Olive Drab over Neutral Gray camouflage. SCAT transports carried tons of cargo and personnel to remote Pacific bases. (USMC)

An SNJ-4 settles in for a three-point landing at Concord, CA during October of 1943. This Texan carried an Intermediate Blue-Gray over Light Gray paint scheme with large Yellow fuselage codes (R-7) just forward of the fuselage national insignia. (Bill Larkins)

This Curtiss SB2C-1C Helldiver of VB-20 carried the large White code "V35" on the nose, identifying the aircraft as being part of a stateside training unit during December of 1943. (National Archives)

An F4U-1 "Birdcage" Corsair (BuNo 02333) of VF-17, the Jolly Rogers, stretches out the wire aboard the USS BUNKER HILL on 11 July 1943. CDR Tom Blackburn, skipper of VF-17, led the squadron to a kill record of 127 confirmed air-to-air kills. (National Archives)

A Grumman J2F-5 Duck of VS-64 shares the sandy ramp at Halavo Seaplane Base, Solomon Islands with PBY-5 Catalinas and an OS2U Kingfisher on 12 July 1943. This seaplane base provided living accommodations for Navy and Marine personnel as well as maintenance facilities for the aircraft. (National Archives)

SBD-3s share the flight deck with F4F-4 Wildcats and TBF-1 Avengers aboard USS ENTERPRISE (CV-6) as she turns into the wind in preparation for launch on 2 May 1942. The ship following in her wake is the battleship USS WASHINGTON (BB 56). (National Archives)

This SNJ Texan and Curtiss SBC-4 Helldiver of the First Marine Aircraft Wing both carried thirteen Red and White rudder stripes during 1941. The SBC-4 was soon phased out in favor of faster monoplanes like the SBD. (National Archives via Dave Ostrowski)

A Vought SB2U-3 Vindicator of VMS-131 on the ramp at MCAS Quantico, VA during early 1942. Vindicators served the USN and USMC during the early part of the Second World War and were phased out shortly after the Battle at Midway. (National Archives via Stan Piet)

This Douglas SBD-3 Dauntless was assigned to VS-5 aboard the USS YORKTOWN during 1941. The Dauntless was powered by a 1,200 hp Wright R-1820 engine and had a top speed of 240 mph and was armed with two .50 caliber machine guns in the nose and a pair of .30 caliber guns in the rear cockpit. (Douglas via Harry Gann)

A section of Blue tailed Vought OS2U-1 Kingfishers of VO-3 aboard USS MISSISSIPPI fly a tight formation during early 1941. The lead Kingfisher later served aboard the USS NEW MEXICO until 4 November 41 when it was lost at sea. (Smithsonian Collection via Stan Piet)

This SBD-5 of VS-64 carried the squadron emblem on the forward part of the fuselage beneath the cockpit and the number 10 in White just forward of the horizontal tail. This Dauntless was a part of a flight of SBDs returning to Segi Airstrip, New Guinea during November of 1943 (National Archives)

An SBD-5 Dauntless dive bomber of Bombing Four (VB-4) returns to the USS YORKTOWN (CV 10) after completing its mission during October of 1943. The aircraft carries a Yagi radar antenna under the starboard wing. (National Archives)

This Consolidated PB2Y-2 Coronado seaplane, beached on Funafuti Island during December of 1943, shows evidence of marine growth on the hull. The nose turret was removed when the aircraft was modified to carry cargo and personnel to forward areas in the Pacific under contract with Pan American Airways. (National Archives)

A PBY-5 (BuNo 08380) of VP-71 undergoes maintenance on its starboard engine at Halavo Seaplane Base, Florida Island, Solomons Islands Chain on 25 December 1943. The weathered Catalina carried two depth charges on wing racks just outboard the wing struts. The aircraft number, 4, was carried in Black just under the horizontal tail. (National Archives)

SBD-5s of VB-5 launch from the flight deck of USS YORKTOWN (CV 10) on 12 August 1943. The size of the national insignias on the SBDs was uniform but the White fuselage numbers differ in size from plane to plane within the squadron. (National Archives)

A flight of PV-1s of VB-125 fly anti-submarine patrol off the coast of Natal, Brazil on 27 October 1943. Ventura B-4 has the Red surround national insignia on the fuselage and wing, while the lead aircraft has the Blue surround national insignia. (National Archives)

This Lockheed PV-1 Ventura carried non-standard placement of the national insignia with a total of eight positions during early 1943. The Navy and Marine Corps received over 1,200 Venturas, many of which saw combat duty during the war. (Smithsonian Collection via Stan Piet)

Douglas SBD-3 Dauntless dive bombers line the ramp at the Douglas El Segundo facility during early 1942. The third aircraft in line, BuNo 03245, served with six different squadrons during its career and was stricken from the inventory on 29 April 44. (Douglas via Harry Gann)

This Hall PH-1 flying boat was assigned to coastal patrol and air sea rescue duties. It had an open bow gun position and another gun on the fuselage behind the wings. The aircraft was powered by a pair of 650 hp Wright R-1820 engines giving it a top speed of just over 150 mph. (Smithsonian Collection via Stan Piet)

Early production Vought F4U-1 Corsairs carried the national insignia in six positions during late 1942. The "bent wing bird" served both the Navy and Marines and was operated from land bases until late 1944 when it was finally cleared for carrier operations. (Smithsonian Collection via Stan Piet)

This Sikorsky JRS-1 of VJ-4 was one of fifteen that served with the U.S. Navy. The amphibian was used as a utility transport and was powered by two 750 hp Pratt & Whitney R-1690 engines giving it a top speed of 190 mph. The JRS-1 could seat up to nineteen passengers. (Smithsonian Collection via Dave Ostrowski)

A Martin PBM-3 patrol flying boat on the seaplane ramp at NAS Norfolk, VA during 1943. As soon as crewmen 'buttoned' it up, the Mariner would taxi into the water where the beaching gear would be removed. (Smithsonian Collection via Stan Piet)

A Grumman JRF-4 Goose in flight over the Atlantic off the coast of Long Island during 1943. These amphibians had provisions for carrying two 250 pound bombs or depth charges on underwing racks located just outboard of the engine nacelles. (Smithsonian collection via Stan Piet)

This overall Yellow Piper NE-1 (BuNo 26381) was assigned to Stations Operations at NAS Norfolk, VA during 1943. It was used for utility work and was powered by a Continental 170 engine giving it a top speed of 95 mph. This Piper was accepted by the Navy in May of 1942 and served until 31 MAR 46 when it was retired. (Smithsonian Collection via Stan Piet)

This overall Yellow Naval Aircraft Factory N3N-1 was one of the "Yellow Peril" trainer biplanes assigned to train future Naval Aviators. This floatplane variant of the N3N was on a ramp at NAS Pensacola, FL during 1943. (National Archives via Stan Piet)

A lineup of Vought F4U-1 "Birdcage" Corsairs of Marine Fighter Squadron 222 (VMF-222), the Flying Dueces, at Barakomo Field, Vella La Vella on 15 December 1943. During the Second World War, VMF-222 shot down a total of fifty-three enemy aircraft. The aircraft at the far right is a Marine F4F-3 Wildcat. (National Archives)

An F4U-1 Corsair of VMF-321 Hell's Angels on the crushed coral taxiway of Samoa Island, Tutuila during November of 1943. This "U-bird" was one of the early "birdcage" Corsairs with the full framed canopy. VMF-321 was credited with 39 air-to-air kills during two deployments in the South Pacific. (National Archives)

An Eastern Division of General Motors TBM-1 Avenger of VC-22 on final approach for landing aboard the light carrier USS INDEPENDENCE (CV 22) on 1 May 1943. VC-22 also flew SBD-4 and SBD-5 Dauntless dive bombers. (National Archives)

F6F-3 Hellcats of VF-5 and Avengers of VT-5 prepare to launch from USS YORKTOWN for air strikes on Marcus Island during September of 1943. The wing gun fairings on these Hellcats identify them as early-production models. (National Archives)

A pair of Columbia Aircraft J2F-6 Ducks of VS-64 parked on the beach at Florida Island seaplane base in the Solomons Islands during July of 1943. The Duck in the background is warming up its 900 hp R-1820 radial engine and will shortly be heading out on a patrol mission. (National Archives)

An F6F-3 (BuNo 40198) of VMO-954 just after touchdown at an auxiliary landing field near Concord, CA during October of 1943. This Hellcat carries the standard "3-tone" camouflage paint scheme used by the Navy and Marines and has a large 17 on the fuselage for easy identification. (Bill Larkins)

A Goodyear FG-1 (85) and a Vought F4U-1A (B112) being prepared for the days training flight activities at Marine Corps Air Station El Tora, CA during late 1943. The Corsair in the foreground is a "birdcage" variant while the other Corsair has the clear canopy that gave somewhat better visibility. (USMC via W.F. Gemeinhardt)

An F4U-1 (BuNo 17451) of VMF-214, Pappy Boyington's Blacksheep Squadron, undergoes routine maintenance in a revetment on Munda Airfield, New Georgia Island during September of 1943. The National insignia has had White bars and a Red surround added to it to comply with the latest marking changes. (National Archives)

These Marine TBF-1s on a South Pacific island base during late 1943 have their cockpits protected from the elements by the attachment of canvas canopy covers. The relentless South Pacific climate and abrasive coral sand took its toll on paint, engines and instruments. (USMC)

These Stearman N2S-3s were used as primary trainers and assigned to NAS New Orleans during 1943. The single-engine two-place biplanes were used to train thousands of Naval Aviators across the nation. (Smithsonian Collection via Stan Piet)

This Yellow winged Vought OS2U-2 Kingfisher was assigned as a seaplane trainer at NAS Pensacola, FL during 1943. The aircraft was accepted on 24 January 1941 and retired during March of 1944. (Smithsonian Collection via Dave Ostrowski)

A Consolidated PBY-5A Catalina parked on a snowy ramp in Alaska during 1943. The aircraft boarding ladder is mounted just behind the port gun blister. Catalinas were used for long-range anti-submarine patrol missions. (National Archives via Stan Piet)

A Grumman F6F-3P (D-61) photo reconnaissance variant of the Hellcat was assigned to VMD-354 at MCAS Cherry Point, NC during late 1943 and used to train Marine reconnaissance pilots before going into combat. When the squadron did see its first combat during early 1945, they flew the improved F6F-5P. (William Derby, Jr.)

This TBF-1 Avenger (V34) carried large stateside training markings and was used for carrier qualification flights for Naval Aviators preparing to join the fleet during late 1943. The Grumman Avenger served both the USN and USMC and was well liked by its pilots for being so sturdy. (National Archives)

This General Motors, Eastern Division built FM-2 "Tall Tail" Wildcat carried the Atlantic Gray and White sea-search paint scheme and was used for anti-submarine patrol duty during 1944. The FM-2 carried four wing-mounted .50 caliber machine guns. (J.E.M. Aviation Slides via Larry Davis)

A Martin PBM-3 Mariner on its beaching gear while undergoing engine maintenance at NAS Norfolk, VA in 1944. A large scaffold was needed to reach the Mariner's twin Pratt & Whitney R-2800 engines. The large dome behind the cockpit housed search radar. (Smithsonian Collection via Stan Piet)

This Consolidated PBY-5 carried the White nose marking "4-M-96" and was assigned to training squadron VN4D8-M at NAS Pensacola, FL during early 1946. The three-tone paint scheme was trimmed with Yellow high visibility markings. (William Derby, Jr.)

(Right) A unique lineup of Naval airpower in the Aleutian Islands during 1944. On the ramp are an OS2U Kingfisher, an NH-1 Nightingale, a JRF Goose and several PBM Mariners. The Green bands on the NH-1 identify it as an instrument trainer. (Smithsonian Collection via Stan Piet)

(Below Right) A Curtiss SB2C-5 Helldiver flying over the countryside during a stateside training flight in 1944. The Helldiver was powered by a 1,900 hp Wright R-2600 radial engine and was capable of reaching a top speed of 295 mph. (Smithsonian collection via Stan Piet)

This Vought OS2U-3 was being craned over the seawall into the water of Pensacola Bay at NAS Pensacola, FL during 1944. Once the Kingfisher was afloat, the rear seat gunner would disconnect the hook. (National Archives via Stan Piet)

A section of SB2C-1C Helldivers of VB-15 from the USS ESSEX carry a full bomb load as they fly in formation toward their Japanese target on 10 May 1944. The Curtiss-built SB2C was armed with two wing-mounted 20MM cannons and a pair of .30 caliber machine guns in their rear cockpit. (National Archives)

# 1944

As the third full year of the U. S. participation in the war began, it found the concept of the Fast Carrier Task Force as the prime naval battle force fully in place. This force included the USS SARATOGA and USS ENTERPRISE plus four ESSEX Class carriers and six light carriers. The fast carriers acted as mobile aircraft strike bases and early in the new year they conducted raids on various atolls in the Marshall Island group. These raids took place every day from 29 January thru 6 February 1944. Kwajalein and Eniwetok were major targets during these attacks, followed by two days of devastating attacks on Truk. As a result of this intensive and concentrated bombing, Truk, the major Japanese naval base in the central Pacific, was rendered useless. It was bypassed since there was no need for U. S. forces to invade and occupy the shattered port.

U. S. strategy called for the prime Navy objective in the Pacific war to be the gaining of the largest islands in the Marianas group to serve as air bases for Army Air Force B-29s to raid Japan. This movement began on 15 June 1944 with a naval force taking up position to invade Saipan, which along with Guam and Tinian were the large islands sought by the USAAF for heavy bomber bases from which to hit the Japanese home islands. This invasion fleet action resulted in the Battle of the Philippine Sea when the Japanese surface fleet elected to engage the American forces.

Just four days later, on 19 June, the enemy fleet made an all-out effort to stop Task Force 58 and the result was the greatest carrier

**Vought F4U-1A Corsairs of VMF-321, along with a lone Grumman J2F Duck, fill the flight deck of the escort carrier USS KWAJALEIN (CVE 98) enroute to the combat area in the South Pacific during June of 1944 (Hank Weimer)**

based air battle of the Pacific war, known as the great "Marianas Turkey Shoot." It gained this name due to the enormous losses suffered by the Japanese. American F6F, TBM, SB2C and even a few SBDs had dealt staggering losses to the enemy air fleet and a total of 346 Japanese carrier based planes were shot down. By the end of July, Guam, Saipan and Tinian were secured and the USAAF had bases to accommodate the growing B-29 Superfortress fleet whose goal was the destruction of Japan.

By September, the ever mobile ESSEX Class carrier strike force of ADM "Bull" Halsey was conducting air raids on the Philippines and quickly destroyed the few enemy planes they encountered. On 23 October, the massive Battle of Leyte Gulf began, which raged furiously for three days. Once again the quality of U. S. carrier aircraft and well-trained Naval aviators took their toll of what remained of the Japanese carrier air units. It was at that time, near the end of 1944, that the new threat of enemy Kamikaze tactics brought the potent F4U Corsair aboard the fast carriers. Flown by Marine aviators, VMF-124 and VMF-213 came aboard the USS ESSEX in the final days of 1944 and became the first Corsairs to operate from carriers. The F4Us proved to be an effective interceptor against the suicide attack threat and earned approval from Navy high command to remain aboard carriers thru the end of the Second World War.

On the Atlantic side, 1944 saw the great German U-Boat menace brought under control. Many Navy patrol squadrons, carefully covering the shipping lanes, contained the Nazi submarine threat. The splendid job done by PBYs, PV-1s and PB4Y-1s paid off. Surface ships containing the materials of war could now move more freely and in relative security to their European destinations. In early June, the long awaited invasion of the European mainland was finally launched although Navy Air had only a limited role in that event.

As the year entered the closing months, the Naval aircraft inventory continued to grow with an awesome volume of production pouring from the many contractors across the U. S. There were plenty of aircraft of all types available to the fleet and all its supporting activities. This had been achieved after just three full years of all-out effort by the American aviation industry. The Naval air component was now well armed and ready to begin what would be the final year of the war.

A Marine TBM-3 (113) tucks away its landing gear after lifting off of its South Pacific base during 1944. With a crew of three, the TBM could carry a 2,000 pound bomb load or a torpedo and could reach a top speed of 276 mph. (Roger Besecker Collection)

A PBM-3D (BuNo 48184) of VP-202 taxis toward a seaplane tender for routine maintenance and resupply on 3 February 1944. A crewman is standing by the radome to make the cable hookup that allowed the Mariner to be hoisted out of the water. (National Archives)

This North American PBJ-1 (BuNo 35094), flying near NAS Patuxent River, MD on 29 February 1944, carries the White marking of the Flight Test (FT) division on the nose. The PBJ Mitchell medium bomber was used exclusively by the Marines in Pacific operations. (National Archives)

This Brewster F3A-1 (BuNo 04656) Corsair of VMF-451 blew a tire on landing and nosed over as a result of hard braking at Mojave, CA on 19 April 1944. The large coding 'A53' was in Yellow on the fuselage and in Black on the landing gear door. (National Archives)

The Curtiss R5C-1 Commando, known for its ability to handle large loads, delivers a Piper AE-1 (BuNo 30289) ambulance aircraft to Peleliu Island on 18 November 1944. This Grasshopper would be reassembled in short order and used to transport litter patients. (USMC via Dave Ostrowski)

A North American SNJ Texan trainer on the ramp at NAS Moffett Field, CA during 1944. The port wing outer wing panel was a replacement section being much darker than the rest of the very weathered SNJ. Late in the war, SNJs became the Navy's primary trainer replacing the earlier biplanes. (Smithsonian Collection via Dave Ostrowski)

An overall Yellow Douglas XJD-1 (57991, ex-USAAF 44-35647) was one of two A-26C Invaders evaluated for use by the Navy during May of 1945. Tests were conducted at NAS Norfolk, VA and ultimately, 140 JD-1s were accepted by the Navy and used primarily as target tows. (via Larry Davis)

A Lockheed PV-2 Harpoon conducts a test flight during 1944. Improving on the PV-1 Ventura, Lockheed increased the wingspan, tail area and fuel capacity to produce a long-range patrol bomber. The Harpoon was armed with five .50 caliber nose guns, a dorsal turret with twin .50s and two more .50s in a ventral mount. (Smithsonian Collection via Stan Piet)

This overall Dark Sea Blue Consolidated PBY-5A Catalina on Guam during April of 1945 carried the fuselage code H-40 on the nose in White. This amphibian had a search radar mounted in the teardrop shaped dome just behind the cockpit. (William Derby, Jr.)

This Vought F4U-1A Corsair on the ramp at MCAS Cherry Point, NC during 1945 carried the fuselage code F-295 in White. Many war-weary Corsairs, as well as other combat types, were reworked and used for stateside training duties. (Picciani Aircraft Slides)

This Consolidated PB4Y-2 (BuNo 59491) of VPB-121 at Camp Kearney, CA during August of 1945 was named TAIL CHASER and carried a nude Indian maiden on the fuselage. After the war, this Privateer continued in service, serving with the Training Command at Corpus Christi, Texas. (Lou Darden)

This overall Natural Metal PBY-5 (08310) carried the nose marking 4-M-93 in Black and was assigned to training squadron VN4D8-M at NAS Pensacola, FL during early 1946. This Catalina did not see combat during the war and was stricken from the inventory on 30 January 1949. (William Derby, Jr.)

This Consolidated PB4Y-2 (BuNo 59480) displays a nude running woman, a Black panther and the name *HIPPIN KITTEN II* on the nose. The Erco nose turret installation mounted a pair of .50 caliber machine guns for forward protection. The aircraft was based in the South Pacific during late 1944. (Paul J. McDaniel Collection)

This overall Natural Metal Douglas BD-2 (BuNo 7095) of VX-2 on the ramp at the Naval Torpedo Station, Quonset Point, RI on 19 July 1944 carries an aerial torpedo in the bomb bay that protrudes beneath the fuselage. (Navy via W. E. Scarborough)

This Curtiss SC-1 (BuNo 35567) left the factory as a landplane but was converted by the Navy to the seaplane configuration. The Seahawk had excellent visibility and was powered by a Wright R-1820 engine, giving it a top speed of 313 mph. (Roger Besecker collection)

This Consolidated PB4Y-1 of VB-103 was used for long range anti-submarine patrols over the Bay of Biscay and the approaches to England while based at Dunkeswell, England during September of 1944. The aircraft carries thirty mission markings in Black under the cockpit. (National Archives)

These Marine Douglas SBD-4s of VMSB-231, returning to their Marshall Island base after a bombing mission on 10 June 1944, carry the distinctive "Ace of Spades" unit insignia on the fuselage, forward of the cockpit. (USMC)

This new production Douglas R5D-1 (BuNo 39139) of VR-5 carried a factory-fresh three-tone camouflage scheme of Dark Sea Blue, Intermediate Blue-Gray and White. This Skymaster was used as a long-range transport for the Naval Air Transport Service (NATS) based at NAS Seattle, WA during March of 1944. (National Archives)

The Brewster SB2A-4 Buccaneer scout-bomber was originally destined for use by the Dutch, but was diverted to the Navy and passed to the Marines. The SB2A-4 was a cousin to the Helldiver and was powered by a Wright R-2600 engine. This Buccaneer was flying near NAAS Saint Simons, GA during March of 1944. (National Archives)

A Curtiss R5C-1 Commando with the side number 82 on the ramp at Camp Kearney, CA in September of 1945. During the Second World War, the Marines obtained 160 R5Cs for transport and cargo use. The R5C had a top speed of 269 mph. (Lou Darden)

A flight of Consolidated PB4Y-2s high over the Pacific near Camp Kearney, CA during 1945. Their original Glossy Sea Blue paint had faded from exposure to intense sunlight and the elements. The aircraft in the foreground carried the fuselage code, E-142, on the nose in White. (Lou Darden)

An overall Natural Metal Douglas R5D-4 (BuNo 90412) high in the California sky during a factory test flight in May of 1945. The Black nose number, 368, was the final three digits of the factory number, 27368. (Douglas via Harry Gann)

A Grumman F6F-5P (BuNo 72104) of VMD-354 being warmed-up prior to a photo reconnaissance mission over Guam during June of 1945. These Hellcats were fully armed and could be used as conventional fighters if necessary. The aircraft behind the Hellcat are Grumman TBF Avengers. (William Derby, Jr.)

A flight of three Consolidated PBY-5 Catalina flying boats in formation over the Gulf of Mexico near NAS Pensacola, FL during early 1946. All carry the White 4M code of training squadron VN4D8-M. Two aircraft carry the overall Dark Sea Blue scheme while the third retains the earlier three tone scheme. (William Derby, Jr.)

These Marine PBJ-1Ds have an ASV radome in the nose and Yagi antennas on the forward fuselage under the cockpit. The Mitchells were being transported from their home base at MCAS Cherry Point, NC to the South Pacific combat zone aboard the escort carrier USS NATOMA BAY during August of 1944. (National Archives)

SBDs were a common sight in the South Pacific theater during most of the Second World War and were operated by the Navy and Marine Corps from both carriers and land bases during 1944. One version, the SBD-3A, was made available to the USAAF and flown by them under the designation A-24. (National Archives)

Marine paratroops jump from a Marine R4D-1 Skytrain over MCAS Cherry Point, NC on 4 May 1944. This Skytrain carries the standard three-tone paint scheme, rare since most Marine R4Ds retained the USAAF Olive Drab over Neutral Gray scheme. (USMC)

A Grumman F7F-1 Tigercat (BuNo 80291) conducts carrier suitability trials aboard the USS SHANGRI LA on 15 November 1944. Although the F7F arrived too late to see active combat during the war, a small number of them did arrive on Okinawa just prior to the Japanese surrender. (National Archives)

F4U-1As (BuNos 55941 and 55928) of VMF-211 parked on the newly completed airstrip on Green Island during March of 1944. VMF-211 was credited with ninety-one kills during the Second World War. The aircraft in the background are Royal New Zealand Air Force P-40s. (USMC)

This battle-damaged F6F-3 Hellcat of VF-29 managed to make it back for a safe emergency landing on the USS INTREPID on 25 October 1944. The ability of the Hellcat to sustain damage and still fly endeared it to the Navy and Marine pilots who flew it. (National Archives)

A Douglas SBD-5 Dauntless in the Atlantic sea-search paint scheme of Dark Gray over White. This Dauntless was assigned to an inshore patrol squadron operating near the southeast Atlantic coast in the area of Beaufort, SC during February of 1944. (National Archives)

A Beech SNB-1 (BuNo 39864) takes off from the auxiliary air field at Concord, CA on 23 March 1944. The twin-engined Beech was used by the Navy as a light transport, as well as for multi-engine pilot and navigation training. (Bill Larkins)

This Martin JM-1 (BuNo 66682) was assigned to VMJ-1 and flew target tow duties. The overall Natural Metal Marauder carried the name, *Joe's Banana Boat*, on the nose in Black. Appropriately, it was stationed at NAS Banana River, FL during July of 1944. (National Archives)

A Vought OS2U Kingfisher on final just prior to making a three-point landing at Concord, CA in September of 1944. The stateside code, WI, was applied in Yellow on the fuselage under the rear cockpit. The Kingfisher could be configured as either a landplane or floatplane. (Bill Larkins)

A Grumman F6F-3 Hellcat, Douglas SBD Dauntless and Vought F4U-1A Corsair share the crushed coral ramp on Green Island during February of 1944. The Hellcat is believed to have been from a visiting Navy unit while the SBD and Corsair are Marine aircraft. (USMC)

The flight line at Marine Corps Air Station Cherry Point, NC was full of F3A-1 Corsairs and a pair of F6F-3 Hellcats during November of 1944. The Corsairs were assigned to VMF-522 and the one marked FT-32 was BuNo 11274. Hellcats saw limited use by the Marines and were primarily assigned to the night fighter role. (USMC)

A new production Vultee SNV-2 (BuNo 52479) in flight over the mountains of southern California during early 1944. This factory fresh Valiant was assigned to NAS Pensacola but had a short-lived career, being stricken from the inventory after only six weeks of service. (A.U. Schmidt Collection via Pete Bowers)

PBJ-1Ds of VMB-413, the first USMC squadron to use the Mitchell medium bomber in combat, parked beneath the palm trees on Espiritu Santo, New Hebrides Islands during March of 1944. Most of the Mitchells carry a radome for surface search radar on the fuselage underside. (National Archives)

Navy F6F-3 Hellcats and Marine F4U-1s share the taxiway at Espiritu Santo Island during February of 1944. Hellcat number 2 carried the name, *Battling Bobbie*, on the fuselage under the cockpit in White. (National Archives)

F4U-1A Corsairs from several different Marine fighter squadrons line the packed coral ramp of the Torokina fighter strip while undergoing routine maintenance on 15 March 1944. (USMC)

The "bomb train" moves down the flight line in front of parked Marine SBDs. These Dauntless dive bombers would soon be re-armed and ready to fly another mission out of Majuro Island during March of 1944. (National Archives)

Vought F4U-1A Corsairs of VMF-321, 4th Marine Air Wing, prepare to depart Iwo Jima after escorting the transport carrying ADM Chester Nimitz to the island during March of 1945. Grumman Avengers, an R4D-1 and a PB4Y-1 are also parked on the field. (USMC)

A Goodyear FG-1A (O-27) taxies slowly down between the rows of parked Marine aircraft while heading for the duty runway at MCAS Cherry Point, NC during June of 1944. The FG-1A was powered by a 2,000 hp P&W R-2800 engine. (Clay Jansson)

This R4D-5 (BuNo 17157) of VR-5 on the ramp at NAS Seattle, WA during September of 1944 carried the Naval Air Transport Service emblem directly under the cockpit. The Skytrain could carry up to twenty-seven passengers and a crew of three and was used by the USN, USMC and USAAF. (National Archives)

An overall Yellow Martin JM-1 (BuNo 75189) of VJ-18 warms up on the ramp at NAS Moffett Field, CA on 7 December 1944. The aircraft had a Flat Black anti-glare panel forward of the cockpit and a Red propeller warning stripe on the fuselage. These aircraft were used for the target-towing mission. (National Archives)

This PB2Y-3 Coronado long-range flying boat was assigned to VP-102 and was being launched from the Navy's major Hawaiian patrol plane base at Kaneohe during March of 1944. The four engine flying boat had a maximum speed of 213 mph and carried a crew of ten. (National Archives)

This Piper NE-1 (BuNo 26197) had its landing gear removed and was attached to an airship by means of a retractable "trapeze" mounting at NAS Lakehurst, NJ during March of 1944. The Cub could be released and recovered while the blimp was inflight, similar to the F9Cs used on the USS AKRON and MACON during the 1930s. (National Archives)

This PBJ-1H (BuNo 35277) was used in carrier suitability tests onboard the USS SHANGRI LA to determine how the multi-engine, tricycle landing gear aircraft handled on a carrier deck. This is the only known PBJ-1 that was modified with a tail hook and it made a successful arrested landing aboard USS SHANGRI LA on 15 November 1944. (National Archives)

An OS2U-3 Kingfisher taxies alongside the heavy cruiser USS NEW YORK during recovery operations on 7 June 1944. An unusual style of aircraft markings were applied to the fuselage with a Black number within a Black circle. These markings were unique to observation type aircraft assigned to Atlantic battleships. (National Archives)

A Lockheed PV-1 (BuNo 34625) of VB-139 taxies down the Marston mat runway at Attu, Aleutian Islands on 1 March 1944. This Ventura was modified with a camera installation beneath the nose. (National Archives)

A PB4Y-1 Liberator provided the backdrop for an awards ceremony for VB-108, VB-109 and VD-4 at Eniwetok Island on 4 June 1944. This Consolidated bomber carried four mission marks stenciled under the cockpit. The number 04 on the tail was in Yellow. (National Archives)

An SB2C-1C Helldiver of VB-15 folds its wings after being spotted aboard the USS ESSEX on 10 May 1944. The aircraft side number was carried on the landing gear doors in Black. The L shaped antenna under the wing was a Yagi radar antenna. (National Archives)

An SOC-3 Seagull of VCS-8 is launched from the port catapult aboard the light cruiser USS BROOKLYN (CL-40) on 12 June 1944 The Seagull was literally blown off the ship since a black powder charge was used to propel the launch sled on the catapult. (National Archives)

The PBN-1 Nomad was a modified tall-tail version of the Catalina, produced by the Naval Aircraft Factory in Philadelphia, PA during late 1944. It was powered by two 1,250 hp R-1830 engines giving it a top speed of 186 mph. Only a few of the 156 planes produced served with the Navy and the majority went to the Soviet Union under Lend-lease. (Navy via Bob Searles)

This Beech JRB-1 Expeditor on the ramp at NAS Clinton, OK during February of 1944 had a fairing added to the top of the fuselage to increase crew visibility. The aircraft was used as a photographic platform by both the Navy and Marines. (National Archives)

This F6F-3 Hellcat (BuNo 41343) of OTU VF-2 served as a training aircraft and carried the name *Bouncing Bette* on the fuselage in White. The Hellcat had just made a wheels-up landing at NAS Melbourne, FL on 10 August 1944, but suffered only minor damage. Navy and Marine pilots said that the name Grumman on an airplane was like the name "Sterling on Silver." (National Archives)

The rear seat gunner of this OS2U-3 (BuNo 5733) is ready to attach the sling line from the heavy cruiser USS BALTIMORE. The Kingfisher pilot, LTJG Denver F. Baxter, had just rescued LTJG George M. Blair, an Essex fighter pilot who was shot down over Truk Lagoon on 17 February 1944. (National Archives)

This F6F-5 Hellcat carried large White markings on its Glossy Sea Blue camouflage. The aircraft ran off the runway at NAS Melbourne, FL on 12 November 1944 damaging the port wing and collapsing the port main landing gear leg. (National Archives)

This overall Natural Metal Consolidated RY-3 (BuNo 90200) served the Flight Test Division at NAS Patuxent River, MD during October of 1944. The RY-3 was the transport version of the PB4Y-2 Privateer and was powered by four P&W R-1830 engines giving it a top speed of 230 mph. (National Archives)

F4U-1D Corsairs of VF-84, VMF-221 and VMF-451 participated in raids over the Japanese mainland. White identification markings were used on the forward section of the nose, wings and tail. These Corsairs were about to launch from the USS BUNKER HILL on 19 February 45. (National Archives)

# 1945

Although it was not obvious at the time, January 1945 marked the beginning of the final year of the worldwide conflict that began on 1 September 1939. There was wide belief that the fight with Germany would in fact be finished during 1945, but it was generally felt in the United States that the war with Japan would drag on throughout 1945, with few people willing to predict it would not end before late 1946 at the earliest. Since the war in the Pacific theater was where the vast majority of Naval and Marine air power was committed, it was there that the buildup of aircrews and their needed aircraft continued. U.S. war plans called for both elements to continue pressing the Japanese back up the Pacific to their home islands.

At the very beginning of the new year, a major U. S. Navy task force was gathered at Leyte Gulf. It soon passed through the Suriagaro Strait where a few days later the Japanese unleashed their newest tactic. A Kamikaze attack by aircraft was made on the escort carrier USS OMMANEY BAY which was quickly sunk. The next day another sixteen or more suicide attacks were carried out which resulted in severe damage to several units of the U. S. surface fleet which by then were in position in Lingayen Gulf for the invasion of the Philippines. The desperate new tactic by the enemy was very costly to the American surface fleet and raised new concerns for effective air defense systems.

The next major campaign in the Pacific began on 19 February 1945 with the invasion of the rather small island of Iwo Jima. This island was targeted because of a need for an emergency airfield roughly half-way along the route flown by the USAAF B-29 bombers from their Marianas Islands bases to Japan. Several Navy escort carriers were in place near the island to provide close air support for the Marine invasion forces. The enemy Kamikaze forces struck again just two days after the invasion began with the victim being the carrier USS SARATOGA. She was seriously damaged by the attack while in position some thirty-five miles northwest of Iwo Jima. The oldest carrier in the U. S. Navy survived but had to return to the United States and was out of action for the remainder of the war.

**A Lockheed PV-2 of VPB-148 parked in its revetment on Johnson Island during June of 1945. The aircraft carried the squadron code Y-119 on the nose and tail in Black. The Harpoon was an improved, larger, more powerful version of the PV-1 Ventura (National Archives)**

Iwo Jima was finally secured on 27 March and Army Air Force B-29s immediately started using it as an emergency landing site and it quickly became a savior for many airmen and aircraft.

As the island hopping movements to the North continued, the next major objective was the invasion of Okinawa which began on 1 April 1945. This proved to be the final massive seaborne assault in the Pacific war. There was no delay in the Japanese use of their most effective new tactic as the suicide aircraft attacks on the U. S. invasion fleet off Okinawa began the very next day.

Over the following two months, nearly 3,000 individual Kamikaze flights were launched against the Navy ships deployed in the Okinawa area. As early as 8 April, about eight U. S. Marine F4Us plus seven night fighters were based ashore on Okinawa's Yontan Airfield in an attempt to counter the mounting number of suicide attacks. The Corsairs provided one of the few effective means of intercepting and destroying the deadly enemy airborne attacks. Okinawa, the final island on the long and costly road to Japan, was at last declared secure on 21 June 1945.

For a four-day period beginning on 24 July, heavy air strikes were mounted by U. S. carrier forces against the home islands of Japan with areas around Kobe and Kure being the most heavily attacked. Navy carrier-based air power had reached the very heart of Japan with significant results. The end was growing near and on 15 August 1945, following the two atomic bomb attacks on Japan, CINPAC sent the order to all Navy an Marine units to "cease all offensive operations against Japan," officially ending the fighting in the Pacific theater.

The formal end was acknowledged on Sunday, 2 September 1945, at 0925 local time in Tokyo Bay when an air armada of 450 Navy and Marine carrier based aircraft together with several hundred USAAF aircraft thundered over the surrender-signing ceremonies being conducted aboard the battleship USS MISSOURI which, along with her massive supporting fleet, was anchored in Tokyo Bay. Navy and Marine Air had most certainly earned a very deserved acclaim of 'WELL DONE' from the highest level of Naval Command and from a grateful nation.

As of VJ Day, 2 September 1945, the inventory of Navy and Marine aircraft was listed as a total of 21,814. The all-time peak of Navy-Marine operational aircraft during the Second World War was reported in the 31 May 1945 Monthly Status Report as 24,861, a staggering number when compared to the 7 December 1941 total of only 2,823.

This sharkmouthed Piper AE-1 (BuNo 30228) was utilized as a DDT sprayer to control malaria carrying mosquitoes around Henderson Field, Guadalcanal during March of 1945. A spray-bar assembly was attached under the fuselage behind the landing gear. (National Archives)

This F7F-3 of VMD-254 saw stateside service with the Marines during July of 1945. The Grumman Tigercat had a Gloss Dark Sea Blue finish and carried White codes with borderless national insignia. (National Archives)

The aircraft parked on the flight line at NAS Kaneohe, Territory of Hawaii during 1945 includes a Cessna JRC-1, several Douglas SBDs, four North American SNJs and two Beech SNB-1s. The Expeditors belonged to Air Base Training Unit 2 and have had their gun turrets removed. (National Archives)

An overall Gloss Sea Blue Beech SNB-2C (BuNo 51218) of training squadron VN18D8 home based at Corpus Christi, Texas on the ramp at NAS Atlantic City, NJ during late 1945. It was used for instrument and cross-country navigation training. (T. Stone via Roger Besecker)

This OS2U-3 Kingfisher was configured as a landplane and carried the Atlantic sea search camouflage of Dark Gray over White with Black side numbers. This Vought-built scout-observation aircraft was flying over pine forests near NAS Atlantic City, NJ during 1945. (T. Stone via Roger Besecker)

An Eastern TBM-3D (side number 12) Avenger parked on the ramp at NAS Atlantic City during 1945. The aircraft carries a wing-mounted surface search radome on the starboard wing. The propeller blade shafts were painted White and the aircraft number (12) was stencilled on the wing fold for easy identification. (T. Stone via Roger Besecker)

This PB4Y-1P of VFP-5 carried the number 54 on the vertical stabilizer in White. This photo-Liberator was flying low over the Pacific on 7 March 1945. The aircraft was severely weathered by the salt air and intense heat of the tropics. (National Archives)

A Cessna JRC-1 (BuNo 64456) of CASU-23 flies over the clouds above NAS Atlantic City, NJ during 1945. This Bobcat was painted overall Aluminum with Flat Black anti-glare panels on the nose and on the inside portion of the engine nacelles. (T. Stone via Roger Besecker)

This Beech GB-2 Staggerwing (BuNo 23694) was assigned to Station Operations at NAS Atlantic City during August of 1945 and carried the standard Navy three-tone camouflage scheme with all lettering and numbers in Black. It was powered by an air cooled 450 hp engine and could reach a top speed of 189 mph. (T. Stone via Roger Besecker)

An SNJ-5 (22 - BuNo 51822) and two SNJ-4 Texans from CASU-23 (Carrier Air Service Unit) fly in formation over the New Jersey countryside near NAS Atlantic City during 1945. The Green fuselage and wing band identified these aircraft as instrument trainers. (T. Stone via Roger Besecker)

An overall Natural Metal Douglas R4D-5 (BuNo 50780) of CASU-21 shares the ramp in front of Hangar LP-12 at NAS Norfolk, VA during 1945 with a number of TBF Avengers and F4U Corsiars. (Navy)

The Grumman JRF-4 Goose was powered by two 450 hp P&W R-985 engines and had two underwing racks mounted just outboard of the engine nacelles capable of carrying depth charges or small bombs. The JRF-4 could accommodate up to seven passengers. (T. Stone via Roger Besecker)

This Vought F4U-4 of VBF-94 flew from the USS LEXINGTON during August of 1945. Letter ship identification codes were carried on the wings and tail during the closing months of the war replacing the earlier geometric identification markings. (Earl Neff vis Jim Wiedie)

This F6F-5 Hellcat of VF-45 was lined up on the port catapult of USS SAN JACINTO awaiting the signal to launch on 13 February 1945. White rectangles were applied to the wings and tail to identify these Hellcats as belonging to the USS SAN JACINTO Air Group. (National Archives)

This Curtiss SB2C-5 Helldiver belonged to the 4th Marine Air Wing and was based on the South Pacific atoll of Eniwetok during March of 1945. The aircraft side number, 19, was carried in White on the nose and on the vertical stabilizer. SB2Cs replaced the SBD as the standard fleet dive bomber. (National Archives)

A PB4Y-2 Privateer high over the California coast near Camp Kearney during August of 1945. The area around the side number had been recently repainted with Gloss Dark Sea Blue, while the rest of the finish was badly faded. The White fairings under the nose housed electronic gear. (Lou Darden)

A PB4Y-1P (BuNo 38894) of VD-1 high over the Gulf of Mexico off the Texas coast near NAS Corpus Christi on 2 February 1945. The large White letter-number code, B77, on the rear fuselage was typical of those carried on stateside training aircraft. (National Archives)

This SB2C-5 (BuNo 88359) of VB-95 on the port deck edge elevator of USS BUNKER HILL on 16 October 1945 carries a White 'T' code letter on the nose and tail. 'T' was the designator for USS BUNKER HILL. The pod under the starboard wing housed a surface search radar. (National Archives)

This Martin JM-1 (BuNo 66735) of VMJ-2 on Tinian Island during June of 1945 carried a high visibility overall Yellow paint scheme with a Black Ace-of-Spades with the letter 'K' in it on the tail. The title, Marine, was stenciled in Black just behind the national insignia. (Clay Jansson via Duane Kasulka)

A Lockheed PV-2 Harpoon of VPB-139 taxies down the steel-mat taxiway on Attu, the western-most of the Aleutian Islands, on 8 April 1945. The aircraft carried the number 24 in Black on the nose and on the fuselage over the insignia. It also carried the number 41 in White on the fin. (National Archives)

This Stearman N2S-2 "Yellow Peril" nosed over from hard braking and sat tail-high just off the runway at NAS Melbourne, FL on 6 April 1945. The overall Silver doped trainer had the fuselage national insignia located unusually far to the rear and had all letters and numbers in Black. (National Archives)

Ground crewmen perform maintenance on a Consolidated OY-1 (BuNo 02779) Sentinel on Okinawa on 3 April 1945. This USMC aircraft was used for artillery spotting and light transport missions. It had a 185 hp Lycoming O-435 engine and a top speed of 129 mph. (USMC via Dave Ostrowski)

The Martin XPB2M-1R (BuNo 1520) Mars seaplane prototype moored at Eagle Mountain Lake, TX on 26 April 1945. The 'R' in the designation indicates that it had been modified as a transport since the bomber prototype first flew in the Summer of 1942. The production version of the Mars had a single tail. (National Archives)

This Consolidated PB2Y-5H (BuNo 7061) of VPB-4 had all gun turrets removed and the openings faired over. It was used for transport and cargo work in the Pacific during July of 1945. This variant of the Coronado had four blade props on the inboard engines and three blade props on the outboard engines. (National Archives)

These overall Gloss Dark Sea Blue F4U-1D Corsairs of VMF-113 on Ie Shima Island during June of 1945 carry three long range drop tanks for maximum endurance. Since Marine Corsairs also operated off fleet carriers, these USMC F4Us retained their tailhook installation. (USMC)

One F6F-5N Hellcat of VN(N)-53 was destroyed and several others damaged on the flight deck of the USS SARATOGA after taking a hit from a Japanese bomb on 17 February 1945. The night-fighters were just preparing to launch when the attack came. (National Archives)

A Marine TBM-3E Avenger (112) and an F6F-5P Hellcat (BuNo 72716) of VMD-354 fly in formation over the South Pacific near the island of Iwo Jima during July of 1945. The Gloss Dark Sea Blue finish on the Hellcat is very weathered and it has heavy exhaust streaking on the fuselage. (William Derby, Jr.)

This Goodyear FG-1D (BuNo 13427) of VBF-4, based at NAS Atlantic City, NJ during August of 1945, had the earlier style braced canopy and carried a Brewster bomb rack on the fuselage centerline. The FG-1D model also had provision for eight underwing HVAR rockets. The code 03 on the tail was in Yellow. (T. Stone via Roger Besecker)

This PB4Y-2 (BuNo 59402) of VPB-118 made a successful crash landing on the emergency field at Iwo Jima during March of 1945. The Privateer was named *Modest O' Miss* and carried a shapely nude young lady on the nose. (USMC via Dave Ostrowski)

An F4U-1D (155) of VF-84 just moments away from landing aboard the USS BUNKER HILL on 16 February 1945. This Gloss Dark Sea Blue Corsair carried Yellow nose markings and Yellow numbers just below the cockpit. The top section of the fuselage forward of the windscreen was sprayed in a Flat Dark Sea Blue to act as an anti-glare panel. (National Archives)

This PB4Y-2 (BuNo 59799) was assigned to Operational Training Unit VB-4-1 at NAS Jacksonville, FL during August of 1945. As with most aircraft assigned to stateside training units, the Privateer carried large identification markings. (National Archives)

F6F-5s of VF/VBF-12 turn up aboard the USS RANDOLPH (CV 15) off Ulithi Atoll during May of 1945. These overall Gloss Dark Sea Blue Hellcats had White ailerons and four White stripes on the tail as ship identification markings. (National Archives)

An overall Gloss Dark Sea Blue F6F-5 Hellcat of VF-40 off the escort carrier USS SANTEE (CVE 29) during April of 1945. The Hellcat carried the side number D11 in White under the cockpit and had heavy exhaust stains on both wings. (National Archives)

*BELLE of the PACIFIC*, a PBM-5 Mariner, being hoisted aboard a fleet seaplane tender on 29 August 1945. Beaching gear has been attached so that the aircraft can rest on the deck. Once routine maintenance has been completed, the Mariner will be returned to the sea. (National Archives)

This Gloss Dark Sea Blue TBM-3E of VT-33 was flown by LTJG Don Kagey off the USS SANGAMON (CVE 26) on 27 March 1945. The installation forward of the windscreen was a gun camera and the pod under the starboard wing held a surface search radar unit. (National Archives)

This Marine F6F-5P (BuNo 72736) photo reconnaissance Hellcat of VMD-354 carried a non-standard underwing fuel tank installation consisting of a pair of 110 gallon P-51 drop tanks and a standard Hellcat underfuselage tank during August of 1945. This field modification was done by the Marines to give the photo Hellcat enough fuel to reach Japan and return to Iwo Jima. (William Derby, Jr.)

A Marine TBM-3 (118) of VMTB-242 prepares to head out on an anti-submarine patrol from Motoyama #2 Airstrip on Iwo Jima during March of 1945. The wings were being spread and the rear crew entry hatch had not yet been secured. (USMC)

This Howard GH-3 Nightingale (BuNo 44970) was assigned to Stations Operations at NAS Atlantic City, NJ during late 1945. This high-winger carried an overall Silver dope scheme with a flat Black anti-glare panel forward of the windscreen and Black letters and numbers. (T. Stone via Roger Besecker)

This PBM-5 carried the name *UMBRIAGO III* on the nose, just forward of the national insignia. The Mariner was being hoisted out of the water and would be craned up to the deck of the seaplane tender which was moored just off the coast of Kerama Retto on 30 March 1945. (National Archives)

An overall Gloss Sea Blue PB2Y-5Z (BuNo 07073) of VJR-1 parked on the ramp at Shanghai, China during late 1945. One local modification that was not envisioned by Consolidated's engineers was the Mark-1 clothesline installation between the twin tails. (Pete Bowers via Roger Besecker)

An overall Natural Metal Curtiss R5C-1 (BuNo 39592) of VMR-252 on the ramp on Iwo Jima during July of 1945. The aircraft was a part of the Transport Air Group and carries the TAG emblem on the lower nose. The Curtiss Commando was used to haul record amounts of much-needed cargo to a variety of central Pacific Islands. (William Derby, Jr.)

This island boneyard was an excellent source for spare parts for various aircraft stationed on Iwo Jima during July of 1944. The aircraft in the boneyard include PB4Y-1s, PB4Y-2s and PV-2s. The Privateer on the far right was BuNo 59446. (Williams Derby, Jr.)

F6F-5 Hellcats of VF-3 aboard USS YORKTOWN carry six 5-in high-velocity (HVAR) rockets on underwing rocket stubs for an air strike against the Japanese home islands on 17 February 1945. The diagonal stripe on the tail and all numbers are in White. (National Archives)

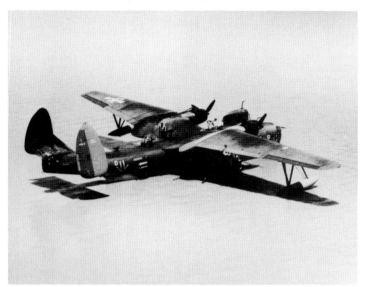

A PBM-5 Mariner (BuNo 59205) of VPB-21 sitting on still water at Chekiang Province, China on 23 July 1945. Salt air and the intense tropical sun has taken its toll on the three-tone camouflage. The aircraft has JATO bottles attached to the rear fuselage. (National Archives)

This overall Natural Metal J2F-6 (BuNo 32640) Duck was attached to Station Operations at NAS Atlantic City, NJ during October of 1945. This Grumman Duck spent its entire service life from the date of acceptance (25 July 1944) until it was stricken (31 March 1946) at NAS Atlantic City. (T. Stone via Roger Besecker)

These Cessna JRC-1s served their military careers with the aircraft pool at NAS Alameda, CA. The Navy acquired forty-seven of these light transports for utility work and declared most of them as surplus at the end of the war. These oil-stained Bobcats are parked, awaiting their fate at San Jose, CA during September of 1945. (Bill Larkins)

An F6F-5 from VF-6 gets the launch signal from the catapult officer aboard the escort carrier USS SUWANNEE (CVE 27) on 20 April 1945. The diagonal identification stripes on the tail, top of the starboard wing and bottom of the port wing were in White. (National Archives)

This Howard NH-1 (BuNo 44926) was assigned to the Naval Air Transport Service and served with VR-2 at NAS Oakland, CA on 15 March 46. It was finished in overall Yellow and carried a wide Green band on the rear fuselage. All letters and numbers were in Black. (Bill Larkins)

An overall Silver doped Beech GB-2 (BuNo 12334) of VRF-3 with the emblem of the Naval Air Transport Service (NATS) painted on its engine cowl. This Staggerwing had a flat Black anti-glare panel and all numbers and letters were also in Black. Some 342 GB-2s were delivered to the Navy. (Bill Larkins)

A PBY-5A Catalina (BuNo 02457) on the ramp at the U.S. Coast Guard Air Station, Brooklyn, NJ. During the Second World War, this Consolidated Catalina logged military service with VP-73, VP-84 and Fleet Air Wing 9 before being stricken on 31 August 46. (T. Stone via Roger Besecker)

This PB4Y-2 Privateer (BuNo 59481), on the coral ramp on Iwo Jima during July of 1945, was assigned to VPB-121 and carried eleven mission markings under the cockpit in White. The very weathered Privateer carried the three-tone camouflage scheme with the nose code, Y481, in Black. (William Derby, Jr.)

An overall Gloss Dark Sea Blue F4U-1D of VBF-6 is maneuvered by plane handlers off the port deck edge elevator to the flight deck in preparation for flight operations from USS HANCOCK (CV 19) on 21 March 1945. The nose ring and tail stripe are in White. (National Archives)

An overall Natural Metal JRF-5 Goose (BuNo 6452) on final approach for NAS Oakland, CA on 1 April 1946. The Goose was assigned to the aircraft pool at NAS Oakland for light transport missions. It carried up to eight, including a two man crew. (Bill Larkins)

A Lockheed R5O-4 (BuNo 12450), assigned to Station Operations at NAS Anacostia on the ramp at NAS Atlantic City, NJ on 8 December 1945. The four stars and anchor emblem on the port engine cowling identify the aircraft as being the personal transport for the Assistant Secretary of the Navy. (T. Stone via Roger Besecker)

A Ryan FR-1 Fireball (BuNo 39703) of VF-41 on the ramp at NAS Oakland on 12 October 1946. Although it did not see combat, the Fireball was in squadron service at the end of the Second World War. The Gloss Sea Blue aircraft had two Yellow stripes on the rear fuselage and a White tail. (Bill Larkins)

This crashed PB4Y-2 (BuNo 59441) of VPB-108 carried the name *Accentuate the Positive* and a nude lady on the nose just to the rear of the national insignia. The Privateer had crashed on Iwo Jima during July of 1945 and was being stripped of usable parts. Two Japanese kill markings were carried beneath the cockpit. (William Derby, Jr.)

A pair of PV-2 Harpoons are parked among rows of USAAF aircraft on the ramp at Andrews AFB, MD a short time after the close of the Second World War. Vast numbers of Navy and Marine aircraft were stricken from the inventory by the end of 1946. (Harold Andrews)

The Grumman XF8F-1 Bearcat prototype (BuNo 90446) was assigned to the Tactical Test Division of NAS Patuxent River, MD during the Summer of 1945. The Bearcat did not see combat during the Second World War since it entered service too late to participate. It was powered by a P&W R-2800 engine and had a top speed of 431 mph. (T. Stone via Roger Besecker)